APPLE TV

THE ULTIMATE BEGINNER'S MANUAL TO USING THE LATEST APPLE TV 4K EASILY WITH TIPS AND TRICKS

BY

FELIX O. COLLINS

Copyright © 2021 FELIX O. COLLINS

All rights reserved. No part of this book shall be reproduced, stored in a retrieval system, or transmitted by any means, electronic, mechanical, photocopying, recording, or otherwise, without written permission from the publisher. Although every precaution has been taken in the preparation of this book, the publisher and author assume no responsibility for errors or omissions. Nor is any liability assumed for damages resulting from the use of the information contained herein.

LEGAL NOTICE:

This book is copyright protected and is only for personal use. This book should not be amended or distributed, sold, quote, or paraphrased without the consent of the author or publisher.

Contents

APPLE TV 4K USER GUIDE

INTRODUCTION

New features of Apple TV tvOS 14

Set up and get started

  Set up Apple TV

    Intro to Apple TV

    Set up an Apple TV

    About Apple TV status indicator

  Set up accounts

    Set your Apple ID on Apple TV

    Set more users on Apple TV

    Set up accounts with other content providers on Apple TV

    With your TV provider, set up a single Apple TV sign-in

    Manage subscriptions on Apple TV

Features of Apple TV 4K

How to use the default password filling on Apple TV and tvOS 14

How to connect an Apple TV to an Apple remote (or Mac)

How to tweak the trackpad on the Siri remote control for Apple TV

Accessibility

  Availability on Apple TV

  Use Zoom to zoom in on an Apple TV

  Set hearing controls on Apple TV

  Install access to Apple TV access shortcuts

  Voice Over

    Use VoiceOver on Apple TV

    Use the braille display with Apple TV

  Switch Control

    About switching controls on Apple TV

    Open "Change Controller" on Apple TV

        Use the control switch on the Apple TV
        On the Apple TV, change the switch control settings
        Scan using the switch on the Apple TV
  Restart, Reset, Update
    Restart Apple TV.
    Reset Apple TV.
    Update Apple TV Software
  Siri and dictation
    Talk to your Apple TV
    On Apple TV, look for entertainment and apps
    Play live content on Apple TV
    Navigate to Apple TV with Siri
    On Apple TV, you can turn Siri on or off.
  Basics
    Use tvOS Control Center on Apple TV
    Customize the Apple TV home screen
    Use iOS or iPadOS Control Center to control Apple TV
    Use the Remote
      Siri Remote overview
      Control what's playing on Apple TV
      Reconnect Siri Remote to Apple TV
      Quickly switch between Apple TV apps
  PODCASTS
    Get podcasts on Apple TV
    On Apple TV, organize podcasts into radio stations
    Control podcast playback on Apple TV
    Adjust podcast settings on Apple TV
  Beyond basics
    Restrict access to Apple TV content
    Family Sharing on Apple TV

Manage Apple TV Storage
Connect a Bluetooth device to Apple TV
Use another remote to control Apple TV
Tips and tricks

INTRODUCTION

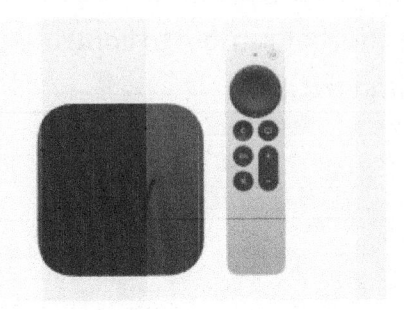

Apple's updated Apple TV is equipped with an A12 bionic processor, 32 to 64GB of storage space, 4K advanced support for HDR support, and a redesigned Siri Remote.

New features of Apple TV tvOS 14

Some of the new features and advancements in the current version of tvOS are listed below. Not all features and information are available in all countries or regions.

Apple TV 4K (second generation) and Siri Remote support (second generation) (tvOS 14.5) The new Siri Remote gives you speed, smoothness, and precise control with the new touchpad. Remote Secrets are clear at a glance.

Instead of speaking, enter Siri queries using the on-screen keyboard or an external Bluetooth keyboard (tvos 14.5), or the iOS or iPadOS keyboard. See typing instead of talking to Siri on Apple TV.

Color balance control (tvOS 14.5) Use iPhone with Face ID and iOS 14.5 or higher to get the best color balance on TV. See the video or audio ratings.

Strength (tvOS 14.3) introduces a new sense of durability powered by the Apple Watch. Sign up for Apple Fitness + for world-class performance on Apple TV. Check out the Apple Fitness app at a glance.

Home Cinema (tvOS 14.2) Connect the HomePod to an Apple TV 4K and listen to music in 4K, surround sound, and Dolby Atmos sound. Install a second HomePod with the most intuitive field and true stereo sound. See Use Apple TV play audio at home.

Multi-user support For multiplayer game support, players can track their game levels, leaderboards, and invitations, and switch between players instantly. Some of your favorite controls, such as the Xbox Elite Wireless Controller Series 2 and Xbox Adaptive Controllers, may be connected to the Apple TV. Check out the Apple Arcade at a glance.

The HomeKit Camera supports receiving door-entry notifications on the screen, checking the front-end view of the front door, and viewing all your HomeKit cameras directly from your Apple TV. Just ask Siri to view any camera at any time, or view all cameras from the main view in the control center. On Apple TV, see How to Use TVOS Control Center.

Controls for audio sharing at home have been improved. Audio sharing allows you to connect two sets of AirPods to Apple TV and watch movies and shows with pals during quiet periods. Enhanced audio capabilities in the Control Center make it easy to play audio on Apple TV from anywhere in your house during party hours. See Use Apple TV play audio at home.

Image-in-picture is now available for all applications that support us. Photo-picture lets you watch movies or shows in a small window while watching games, watching the weather, or having fun on TV shows. See Control playing on Apple TV.

4K Video Sharing Using AirPlay, you can share 4K videos in the "Photos" app on iPhone or iPad on Apple TV 4K with full resolution. See AirPlay allows you to transmit audio and video to your Apple TV.

4K YouTube supports watching the latest YouTube videos in full 4K glory. Your favorite music, slow motion, outdoor videos and vlogged have never been better.

APPLE TV 4K USER GUIDE

Set up and get started

Set up Apple TV

Intro to Apple TV

This document explains how to install tvOS 14.5 on Apple TV 4K and Apple TV HD.

See Set up Apple TV for instructions on how to get your Apple TV up and running.

Set up an Apple TV

To use Apple TV, you need to:

- HD or 4K TV with HDMI
- To connect Apple TV to a television, you'll need an HDMI cable (for 4K HDR, you may need a high-speed HDMI cable, sold separately)
- Access to a wireless 802.11 network (wireless streaming requires 802.11a, g, n, or ac) or Ethernet, as well as high-speed Internet
- Apple ID for buying and renting, downloading apps from the App Store, and using family sharing

You'll need a TV that supports 4K, HDR, and Dolby Vision, as well as a Dolby Atmos compatible audio system, to fully enjoy Apple Audio's 4K audio and video capabilities.

See the Apple Support article on 4K, HDR, and Dolby Vision on

Apple TV 4K for more information on how to set up Apple TV 4K on a 4K TV.

For more information on setting up Apple TV 4K audio surroundings (including Dolby Atmos), see the Apple Support article entitled Playing Audio on Dolby Atmos or Playing Audio Surround on Apple TV.

Apple TV preview, all models:

Rear image of the Apple TV 4K:

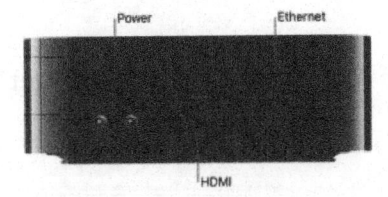

Rear image of the Apple TV HD:

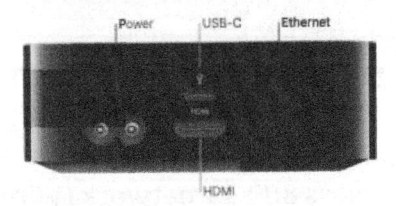

About Apple TV status indicator

The status indicator on the front of the Apple TV shows the following:

If the Apple TV is	the Status lamp
On	Light

APPLE TV 4K USER GUIDE

Off or set aside

Get started

Receive orders from afar

Update software

closed

Blink a little

Turn it on once

Fast light

Set up accounts

Set your Apple ID on Apple TV

Your Apple ID is the account you use to buy movies and TV series on Apple TV, as well as subscribe to Apple TV channels in the Apple TV TV app. and downloading apps in the App Store. You can also use iCloud, which connects you to all your Apple devices to share photos and more.

If you already have an Apple ID, install it when you first set up an Apple TV. If you do not have an Apple ID, you can create it on the Apple ID web page. You only need one Apple ID to use the Apple TV, iTunes, Cloud, and Game Center app. Visit the Apple ID FAQ page for additional information.

Here are a few things you can do with your Apple ID on an Apple TV:

- Apple TV app: Buy or rent movies, buy TV episodes and episodes, and subscribe to Apple TV + or Apple TV channels within the app. You can also use your Apple ID to access purchases made on other devices. Music: If you are a subscriber to Apple Music, you can access millions of songs on Apple TV. And with Apple Music subscriptions (or iTunes Match subscriptions purchased separately), you can access all of your music on all your devices - including music you downloaded from CDs or purchased elsewhere outside the iTunes Store. Check out the Apple Music app at a glance.
- Apps and Arcade: Buy apps or subscribe to the Apple Ar-

- cade directly on Apple TV and download previous App Store purchases on Apple TV for free, anytime.
- Eligibility: Use with your Apple Watch and Apple Fitness+ subscriptions, which include a world-class performance by top trainers. See the Apple Fitness app at a glance.
- Game Center: Play your favorite games with friends with Apple TV, iOS or iPadOS, or Mac (OS X 10.8 or later).
- Photos: View your photos and videos from Cloud Photos, My Photo Stream, and Shared Albums. Check out the Apple TV photo app at a glance.
- Family Sharing: Share purchased movies, TV shows, apps, and subscriptions on your Apple TV with up to six family members. Watch Family Sharing on Apple TV.
- One Home Screen: Keep your installed apps and home screen look the same on every Apple TV you have. See Customizing the Apple TV Home screen.
- Sign in with Apple: Sign in to apps with the Apple ID you already have, without having to fill out forms or create new passwords. Apple does not track your activity and control your data.
- Support for AirPods: Listen with AirPods without having to set them up. AirPods connected to your Apple ID automatically connect to your Apple TV.

Please keep in mind that not all features are available in every country or region.

Set more users on Apple TV

You can share your Apple TV with several family members so that they all have access to your content preferences and Apple TV App, Apple Music, and App Store accounts. After adding family members to your Apple TV, you and your family can switch between account profiles to create a personalized view.

Please keep in mind that not all features are available in every country or region.

Set up accounts with other content providers on Apple TV

Some apps and content providers (like Netflix and HBO Max) require you to log in before you can use them. When you open an application to sign in or set up an account, follow the instructions on the screen.

Tip: You can specify your username and password. After the text input field is displayed on the screen, press and hold the Siri button on the remote control button, then speak letters, numbers, and symbols instead of using the on-screen keyboard. It says "uppercase" to indicate capital letters. Use Siri Dictation on Apple TV for further details.

With your TV provider, set up a single Apple TV sign-in

Single login makes it easy to access entertainment apps such as HBO Max and FXNOW. You just need to log in to Apple TV once to have access to all supported applications that require cable or satellite pay-TV authentication.

Note that some cable and Internet companies provide zero-login options. which further simplifies the login process by default login. During the first setup, follow the on-screen directions.

You can use single sign-in during Apple TV setup or settings. After logging in, any other supported apps that you can access will verify you without logging in again.

Tip: If you store the password on an iOS device with iOS 12 or later or with iPadOS with iPadOS 13 or later, you can automatically fill in the account password on Apple TV. See Set up accounts with other content providers on Apple TV.

Activate or disable single login

1. Open settings on Apple TV.
2. Go to users and Accounts> TV Providers and select your

TV provider
3. Sign in with your provider username and password.
To disable one login after login, select opt-out.

Note: If you have multiple Apple TVs, you must sign in once for each Apple TV to use this feature. Single sign-in is available for iOS and iPadOS.

Manage subscriptions on Apple TV

Apple TV subscriptions, such as Apple Music, Apple TV channels, Apple Arcade, and individual apps, can be modified or canceled.

Manage subscriptions

1. Open settings on Apple TV.
2. Go to Users and Accounts> [Account Name] Subscriptions and select Subscriptions.
3. Follow the on-screen instructions to change or cancel your subscription.

Please keep in mind that not all features are available in every country or region.

Features of Apple TV 4K
Reviews

The review of the second-generation Apple TV 4K is a lot better, with reviewers recommending a faster A12 chip and a redesigned Siri Remote. The following review video shows a complete understanding of Apple TV 4K and its new remote control.

The new Apple TV is said to be "extremely fast" and offers better information than ever before. The reviewer believes the A12 chip is faster but said "there is no difference between day and night." The main advantage of the new chip and the new support for HDMI 2.1 is the ability to play HDR content at a high speed of 60 frames per second.

Currently, the amount of content that supports 60 frames per second is limited, and users need compatible TVs. Some commentators believe that content is 60 frames per second, like sportswear, smooth, but invisible to the untrained eye.

The new Siri Remote has a brand new design and a five-way touchpad. Critics have praised this as a bigger improvement than the previous model.

While there is general admiration for the new remote control design, there is also criticism of the back button. Because different developers use it in different ways, its performance is much more structured than expected.

Some critics have pointed out that compared to other broadcast media outlets and sticks, the Apple TV is still more expensive, while others say it offers a "more reliable platform" to prove the price.

Comments can help potential buyers decide to buy the new Apple TV 4K. More details can be found in our dedicated review.

Hardware and design

Apple TV 4K did not receive design updates when it was updated in 2021. It is still a simple black, invisible box, the size of a palm tree. Its four sides are 3.9 inches long and 1.4 inches high. In terms of weight, 15 ounces is less than a pound.

There is an Apple TV logo on top. In addition, this device is black on all sides, blending well with home decor. It is small enough to fit on any shelf near the TV or TV.

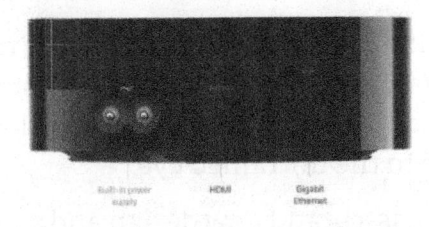

Behind the Apple TV 4K, there is an HDMI 2.1 port, a Gigabit Ethernet port, and a power plug point.

Processor and internal design

The second-generation Apple TV 4K is fitted with an A12 bionic chip, similar to the chip used for the first time on the iPhone XS, XS Max, and XR in 2018.

The A12 Bionic chip is a major development of the A10X Fusion chip in the previous generation of Apple TV 4K, which improves CPU and GPU speeds.

To store games and downloaded content, Apple TV 4K offers 32GB or 64GB capacity.

Communication

Apple TV 4K supports 802.11ax WiFi 6. WiFi 6 is the latest and fastest WiFi agreement, with high speed, extended network capacity, low latency, high power efficiency, and improved performance in areas with many smart home devices.

The second-generation Apple TV 4K is the first Apple TV with built-in Thread support, in addition to WiFi 6. Thread is a low-power network technology that allows for secure communication, a grid-based system that can work with other smart home devices that support Thread to improve communication.

The Apple TV 4K also supports Bluetooth 5.0 and has a remote eye input receiver.

4K and HDR

The second-generation Apple TV 4K, like the previous Apple TV 4K, offers 4K HDR video, which provides brighter, more brilliant colors and greater clarity than the 1080p Apple TV HD.

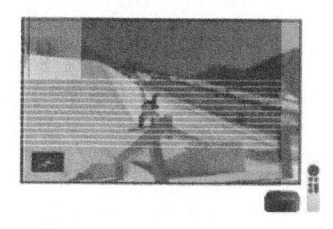

It supports HDR10 and Dolby Vision simultaneously, and the new version of the second generation supports high-quality HDR, bringing smooth and clear videos. Thanks to the HDMI 2.1 port, the second-generation Apple TV 4K offers twice the rate of high-resolution video frames, up to 60 frames per second.

Apple said the fast-moving movement looks smooth, the natural recordings are bright, and YouTube videos are "off the screen."

To view material in 4K quality, Apple TV 4K requires a compatible 4K TV, and features like Dolby Vision require a TV that supports Dolby Vision. Apple TV 4K uses HDMI to connect to a TV. HDMI cable not installed.

4K streaming

To watch 4K material from iTunes, Netflix, or other sources, you'll need a 4K capable device, Apple recommends that customers have a connection speed of at least 25Mb / s. If the internet connection is not fast enough to transfer 4K content, Apple will reduce video quality.

Apple does not allow users to download 4K content from iTunes, and 4K content is limited to streaming media playback.

Image, video, and video formats supported

H.264, HEVC (H.265), HEVC Dolby Vision, and MPEG-4 are all supported by Apple TV 4K. Images in the following formats can be displayed: HEIF, JPEG, GIF, and TIFF. HE-AAC (V1) and AAC audio formats are supported (up to 320 Kbps), protected AAC (from the iTunes Store), MP3 (up to 320 Kbps), MP3 VBR, Apple Lossless, FLAC, AIFF, and WAV (Dolby Digital 5.1), E-AC-3 (Dolby Digital and 7.1 surround sound), and Dolby Atmos (Dolby Digital and 7.1 surround sound).

Apple TV supports local audio via Dolby Atmos and standard non-lost audio, from 16 bits at 44.1 kHz to 24 bits at 48 kHz. Apple said Apple TV 4K "currently does not support Hi-Res Lossless", which opens the door to future software updates and will provide Hi-Res Lossless device support in the future.

Color balance

In addition to the Apple TV 4K in 2021, Apple also added a new color measurement feature that uses an iPhone with a face ID camera to measure colors on an Apple TV.

- How to use the Apple TV-based balance function

In the video section of Apple TV settings, there is a color balancing option to enable this feature. You can hold your iPhone near the TV screen, and the iPhone will find the color and compare it to the industry standards to make adjustments. This function is not available when Dolby Vision is enabled for compatible TV.

Siri remote control

Apple has introduced the redesigned Siri Remote to match the second-generation Apple TV 4K. The new remote is large and uses an aluminum body. The updated Siri Remote is equipped with a touch track pad that supports gestures instead of high touch.

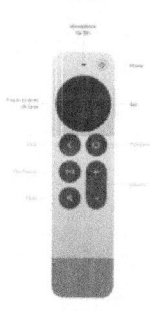

One-click allows you to enter a TV game or move, and swipe to get through a long list of content. You can slide the touchpad exterior to browse shows and movies, fast forward, and rewind when needed. Remote includes TV/home button, back button, play/pause button, silence button, and set of volume buttons.

A new remote control function is a dedicated power switch or TV switch. On the remote control side, there is a dedicated button to activate Siri. Siri's performance is limited to certain countries/regions, and the list is provided on Apple's website.

In countries/regions that do not support Siri support, Siri Remote is called "Apple TV Remote."

Siri on Apple TV is very similar to Siri on iPhone. You can talk to the remote, and Siri commands will be restored to the Apple TV. Siri can be used

There is a Lightning hole at the bottom of the Siri Remote, which can be charged every few months. The previous version of the Siri remote includes an accelerometer and gyroscope, which can be used as an Apple TV game controller, but the new remote does not have this feature. Instead, Apple wants users to use a Bluetooth controller to play games.

Siri Remote does not have the rumored Find My Link feature, but people have been trying to use AirTags as another way to track.

- Apple TV: How to customize trackpad for second-generation Siri remotely
- Apple TV: How to use the second generation Siri Remote to browse videos

Remote request

In addition to the visual Siri Remote, Apple TV can also be controlled using the Remote app provided in the App Store for iPhone, iPad, and Apple Watch. Remote application configuration is similar to Siri Remote, providing visual navigation controls i-Apple TV interface, Siri access, and volume control.

Bluetooth accessories

Popular controllers from PlayStation, Xbox, and Nintendo can be paired with Apple TV for gaming purposes. Apple hopes that Apple Arcade and other games can be played using a controller instead of Siri Remote.

The Apple TV 4K can also be paired with a Bluetooth keyboard, but you can use an iPhone tied to the same iCloud account to search and type for other purposes.

TvOS and TV apps

TvOS is an active program that runs on Apple TV 4K and Apple TV HD and provides an easy way to watch TV views on Apple's set-top boxes.

TvOS has a complete App Store, supports downloading a series of various apps and games that can be used on Apple TV, and interface design sets original content. Use Siri commands, Apple Remote or Remote app on iPhone and Apple Watch to find what you're looking for, and track the app with "Up Next".

TvOS includes many built-in programs, such as your photo library access, Apple Music, Podcasts, and Apple TV, which can integrate TV and movie content from multiple sources, including Apple TV streaming services +.

Channels are included in the Apple TV app, so you don't need to open a third-party app to subscribe and watch content from paid services. Apple will make suggestions for content you'd like to watch. You can use the home app to manage your smart home products.

Apple will always be adding new features to tvOS.The update to tvOS 2021 is tvOS 15. It will be a fan of tvOS 14 and will be launched in the fall. TvOS updates will never bring as many changes as iOS or macOS updates, but there are still some new features that are noticeable on tvOS 15 that you should highlight.

FaceTime's SharePlay function allows many users to watch TV shows or movies at the same time, It will be integrated with tvOS, and a new engine will make recommendations. "For All of You" suggests programs that everyone might like.

The Shared Section also shows movies and programs shared with you through the messaging app, so remember to check them out. When paired with AirPods Pro or AirPods Max, Apple TV supports Spatial audio, provides a cinematic sound experience around the cinema, and has a new feature that automatically connects to your AirPod using the Smart AirPods channel.

Two HomePod mini speakers can be paired with Apple TV 4K stereo. If you have a HomeKit camera, you can watch multiple cameras on an Apple TV at the same time.

How to use the default password filling on Apple TV and tvOS 14

If you lock between characters on the on-screen keyboard, or call numbers and numbers in Siri Remote, accessing apps and services on Apple TV can be frustrating. To make it easier to enter login details, Apple introduced the Continuity Keyboard, an iOS feature that allows users to enter text on Apple TV using an iPhone or iPad.

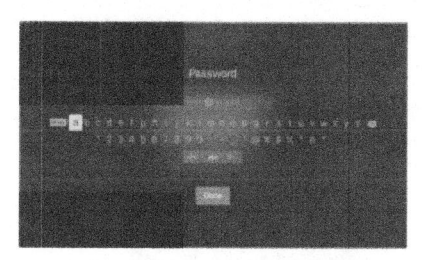

In tvOS 12 introduced this fall, the continuation keyboard has been further upgraded to support automatic password filling. So, whenever you encounter a login screen, you will see the following notification on your iPhone or iPad.

To bring up the keyboard, simply tap the warning at the top of the screen. and on the Apple TV, you should be able to fill in your password information by tapping AutoFill in the quick type column. The same basic login method applies if you're using an Apple TV remote or control center.

The above assumes that your Apple TV and iOS device is on the same iCloud account. But suppose you are a guest in someone's home and you want to sign in to your Apple TV account. What then?

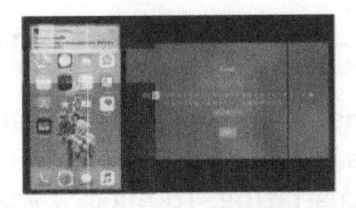

Thankfully, tvOS 14 is also well suited to this situation. Apple TV can now hold Siri Remote to find nearby iPhones. After finding your device, your iPhone will ask you to confirm if you want to use the default charger. It will prompt you to enter the verification PIN displayed on the nearest Apple TV.

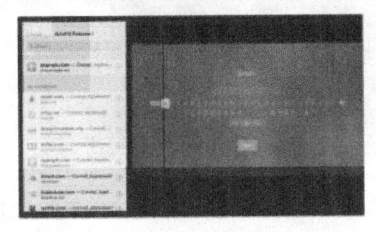

If you use Face ID, Touch ID, or password to verify your iPhone, you will see a list of passwords, and the passwords of related apps or services you want to access will be displayed at the top, ready to send the Tap to your friends on Apple TV.

The automatic filling of the password on Apple TV requires tvOS 12 and iOS 14, both of which will be released in the fall.

How to connect an Apple TV to an Apple remote (or Mac)

When you set up a new Apple TV and start a set-top box, as long

APPLE TV 4K USER GUIDE

as you press one of the buttons, the Apple Remote in the set-top box will be paired automatically. If Apple Remote stops working, it may run out of power and need to be charged via USB to a Lightning cable connected to a USB socket for 30 minutes.

But if this doesn't solve the problem, the best way to pair the device with Apple TV. This article tells you how. If you need to pair a new Apple Remote site in case the Apple Remote that comes with Apple TV completely stops working or is irreparably damaged, the following instructions will also be very helpful.

Additionally, at the end of this article, we provide a quick tip for pairing Mac and Apple TV Remote to control content such as iTunes, VLC, and Keynote.

How to connect an Apple TV to an Apple remote

1. Ensure that your Apple TV is switched on.
2. Point the Apple Remote 3 inches away from the set-top box, then press and hold the remote control buttons "Menu" and "Volume Up" for 5 seconds.
3. If you see a notification on the TV screen asking you to go to Apple Remote, place the Remote on top of the Apple TV.
4. If you do not see a remote connection notification on the TV screen, unplug the Apple TV from the wall store, wait at least six seconds, and then reconnect it.
5. If necessary, repeat steps 1 to 3.

How to use Apple's remote to control your Mac

Apple once equipped other Macs with a small white or silver infrared control, allowing Mac users to control Keynote presentations and iTunes media remotely.

When new Macs no longer include IR receivers (indicated by a black line on the front edge of the fuselage), Apple stopped installing these remotes, but Apple TV users may choose to use their Apple TV Remote to control their Mac. Thanks to free third-party Eternal Storms Software's party macOS utility called SiriMote.

SiriMote is not available in the App Store, but you can download it directly from the Eternal Storms [Direct Link] website. Drag SiriMote from the "Downloads" folder to the "Applications" folder, then open the software and follow the directions on the Apple Remote pairing screen with your Mac.

APPLE TV 4K USER GUIDE

How to tweak the trackpad on the Siri remote control for Apple TV

The second-generation Siri Remote has a portable trackpad for navigating menus and accelerating and rewinding films.

When you start using Siri Remote for the first time, the location touch is turned on automatically, but if you can't maintain sensitivity or are not used to using the touch control on the remote, you can customize it or switch it on at any time.

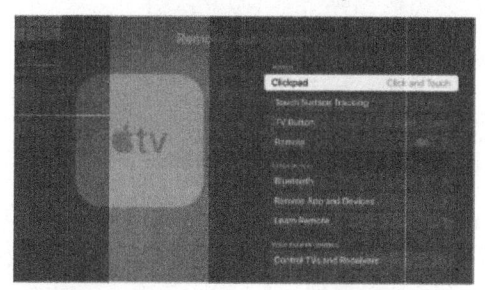

Change Siri remote touch pad function

1. Go to the Settings app on Apple TV.
2. Select remote and device.
3. Choose a fingerprint clip.
4. Select "Click Touch" to enable touch and touch location tracking, or select "Click Only" to turn off touch location tracking.

FELIX O. COLLINS

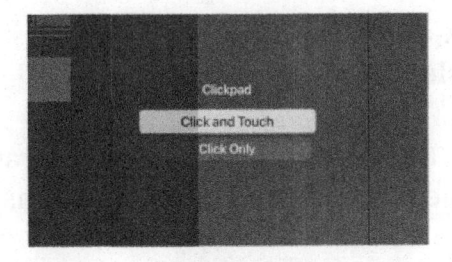

Change Siri remote-tracking sensitivity

1. Go to the Settings app on Apple TV.
2. Select remote and device.
3. Select Touch location tracking.
4. Select "Fast" to make small thumb movements further on the "Apple TV" screen, or select "Slow" to reduce tracking sensitivity.

If you are having trouble browsing the video while the play is paused, please check our little guide to make sure you are using the right touch of the trackpad.

Accessibility

Availability on Apple TV

Apple TV features built-in access features:

- Narrative: Apple TV supports the narrative, Apple screen reader. To support all Apple TV-supported languages and braille displays, VoiceOver can tell you exactly what is on the TV screen and help you choose commands. See Use VoiceOver on Apple TV.
- Zoom: Zoom built-in magnifying glass that can be used anywhere on the Apple TV screen. Zoom enlargement can be changed up to 15 times in standard size, which can help with a series of vision challenges. See Use zoom to zoom in on Apple TV.
- Display adjustment: Change the parameters on the Apple TV to achieve color contrast, light sensitivity, and brightness. See Use Adjusted Editing on Apple TV
- Bright Text: Select bold text to make text readable throughout the Apple TV interface. See Display bright text on Apple TV.
- Increase brightness: Increase contrast on-screen by reducing background visibility on movie and TV show pages, menu tabs, etc. You can also use top-notch indicators to better display important content. See Increasing Apple TV screen variability.
- Reduce fitness: By slowing down exercise, certain screen functionality (like moving between app icons on the home screen and launching apps) looks easier. See Reduce screen movement on Apple TV.
- Audio Definition: Audio Definition provides descriptions of important screen performance and content on movies and TV. See the description of using audio on Apple TV.
- Change controller: Use a connected Bluetooth® device or another platform such as switching controls to Apple TV

See About changing controls on Apple TV.
- Apple TV features closed captioning and SDH, allowing deaf and hard-of-hearing users to fully appreciate the new content. episodes and thousands of movies. When browsing movies or TV shows in the iTunes Store, simply look for the CC or SDH icon. You can customize subtitles with special styles and fonts. See Use subtitles on Apple TV.
- Type Siri: Use an on-screen keyboard on Apple TV, Bluetooth wireless keyboard, or iOS or iPadOS device to type Siri queries. See typing instead of talking to Siri on Apple TV.
- Siri: Siri Remote lets you pass screen roaming using voice. Let Siri do a variety of activities, such as "find children's movies", "take back five minutes", "turn on music", and more. See Talk to Apple TV.

When the on-screen keyboard appears, highlight the text input field and press and hold the Siri button. Specify text or spell username and password instead of typing. See Use Six Dictation on Apple TV.

Use Zoom to zoom in on an Apple TV

Zoom in and pan around the zoom image using the touch area on the remote.

Turn Zoom on or off

- In the Apple TV settings "Settings", go to "Accessibility">"Zoom" and open "Zoom".

You can also set accessibility shortcuts to activate zoom. See Add Apple shortcuts to Apple TV.

Zoom in or out

- Press the center of the touch pad (second generation Siri Remote) or the touch area (first generation Siri Remote) three times to activate Zoom.

Anything you highlight will be magnified automatically.

APPLE TV 4K USER GUIDE

Move the Zoom focus

- Tap up, down, left, or right edge of the touchpad (Siri Remote 2nd Generation) or touch area (Siri Remote 1st Generation) to move the screen object in that direction.

Adjust the magnification

- Tap the touch pad (second generation Siri Remote) or touch area (first generation Siri Remote) and drag it up or down with two fingers.

To limit maximum magnification, go to "Accessibility"> "Zoom"> "Maximum zoom level" in Apple TV "Settings" settings.

Turn panning on or off

- Tap the touch pad (second generation Siri Remote) or touch pad (first generation Siri Remote) with two fingers.

Pan to see more

- When swinging, drag your finger on the touch pad (second generation Siri Remote) or touch area (first generation Siri Remote).

Speak the currently selected screen item

- Press the Siri button twice in Siri.

Set hearing controls on Apple TV

You can set hearing controls so that Apple TV can only output mono audio. You can also change the audio balance between left and right speakers.

Turn on mono audio

- In the Apple TV "Settings" settings, go to "Accessibility" and open "Mono Audio".

Adjust audio Balance

- In Apple TV's "Settings" settings, go to "Accessibility", then select "Balance" and adjust your slide controls.

Install access to Apple TV access shortcuts

You can add access shortcuts to the back button (second generation Siri Remote) or menu button (first generation Siri Remote). If you press the back button or menu button three times, the shortcut will start.

Add an accessibility shortcut

- In the Apple TV settings "Settings", go to "Accessibility"> "Accessibility shortcuts" and select an option.

Use your accessibility shortcut

- In Siri Remote, press the back button three times to return to the button (menu button in the first generation Siri Remote).

Voice Over

Use VoiceOver on Apple TV

VoiceOver lets you control Apple TV without looking at the screen. To traverse the screen, simply touch the Touch surface of the Siri Remote and listen to VoiceOver talk to each item you highlight.

VoiceOver has two options:

- Navigation: Speak selected items on the screen as you navigate with Siri Remote.
- Check: Save the currently selected item so you can use Siri Remote to browse the screen. This allows you to hear objects and text elsewhere on the screen before selecting.

Turn voiceover on or off

- In the Apple TV settings "Settings", go to "Accessibility"> "Account" (or use accessibility shortcut).

See Add Apple shortcuts to Apple TV.
Ask Siri. Say this:
- "Open account"
- "Close account"

Switch between navigation and exploration modes

- Use two fingers to tap three times in the center of the touch pad (second generation Siri Remote) or the touch area (first generation Siri Remote).

Use voiceover in navigation mode

When you drag your finger between the trackpad area (Siri Remote second generation) or touch area (first generation Siri Remote), VoiceOver will speak to each menu or text item of your choice.

1. To move your selection, scroll up, down, left, or right.
2. To select an item, press the center of the trackpad or touch more.

Activate voiceover rotor controls in navigation mode

The VoiceOver Rotor allows you to use gestures in the middle of the trackpad (second generation Siri Remote) or in the touch area (first generation Siri Remote) to adjust the VoiceOver controls, such as speech speed.

Tip: Reading rotor controls on the screen after a delay allows you to hear any text items left on the screen, such as movie descriptions or cast lists.

1. Rotate two fingers in the center of the track or point in the touch area to open the VoiceOver Rotor with additional controls.

Available rotor controls vary depending on your performance. When you turn on the rotor, you will hear a sound before VoiceOver speaks the currently selected controller.

2. Do any of the following:

- Switch to another control: Rotate two fingers to the left or right again.
- Adjust controls: Swipe up or down. When you swipe, VoiceOver will speak the currently selected number (for example, "55%").

For example, to adjust speech volume, rotate two fingers until VoiceOver speech level control is selected. Then swipe up or down to increase or decrease the level of speech.

- Customize rotor function: Rotate two fingers until the custom function is selected on the Rotor. Swipe up or down to hear each custom action, then press in the middle of the trackpad or tap up to act.

3. To exit the Rotor, swipe left or right, or wait about three seconds until you hear a sound coming out.

Use voiceover in exploration mode

By default, when you browse nearby objects or other objects on the screen, the object selected in test mode is always selected. This lets you take a closer look at what's available on the screen.

When you drag your finger on the trackpad (second generation Siri Remote) or touch area (first generation Siri Remote), VoiceOver highlighting also speaks to every menu or text item you browse.

Do any of the following:

- Hear the name of the item: Swipe left or right. Item is also highlighted on the screen.
- Select highlighted item: Press in the middle of the trackpad or touch more.

The newly selected item is always selected until you browse and select another item.

- Adjust the currently selected VoiceOver controller: Swipe up or down. See activity below.
- Speak everything from the current text object to the bottom of the screen: Swipe down with two fingers.

- Read items on the screen from the top: Swipe up with two fingers, or press and hold the play button.
- Pause or resume speaking: Tap once with two fingers to pause. Tap again with two fingers to recover.

Use the braille display with Apple TV

When VoiceOver is turned on, you may read VoiceOver output and control the Apple TV using the Bluetooth® braille display using a braille display with input buttons and other functions. See the Apple Braille Supporting Articles for iPhone, iPad, and iPod touch braille for a list of approved braille advertising.

Connect a braille display

1. In the Apple TV settings "Settings", go to "Accessibility"> "Accounts"> "Brille".
2. Choose a braille display to pair the display with an Apple TV.

When VoiceOver is used with a braille display, the display will print a screen text of the object you are hovering over. When you move focus, VoiceOver will speak and the braille display will print text. The braille display can include additional buttons that support Apple TV's basic navigation.

Adjust braille settings

- In Apple TV settings, go to "Accessibility"> "Accounts"> "Braille", and then do any of the following:
  - Select "Submit" or "Import", then select "Reduce", "Unshrinked eight points" or "Remove six points".
  - Choose the default translation, then choose to close or close.
  - Select a braille table, then select a table or choose to add a braille table.
  - Select the duration of the notification display, then swipe to select the duration.

During media playback, output braille closed captions

- In the Apple TV settings "Settings", go to "Accessibility"> "Detailed Level"> "Media Description", then select "Braille" or "Talk with Braille."

Set the language for voiceover

- In Apple TV settings, go to "General"> "Language and region" and select Language and Regional Options.

If you change the language of your Apple TV, you may need to reset the VoiceOver language and braille displays.

A more detailed description of the status cell may be found here

- In your braille display, press the Route button on the status unit.

Switch Control

About switching controls on Apple TV

Switch assistive technology that allows you to control the Apple TV by tapping the keyboard, mouse button, or other controls on a connected Bluetooth® device as a switch. Use any variety of task modes such as selecting, dragging, and typing. You can use the switch to select an item on the screen, and then use the same (or different) button to select the action you want to perform on that item or location. The basic method is:

- The scanner (default), will highlight various items on the screen until you select one
- The manual selection allows you to move from project to project as needed (multiple changes required)

No matter which method you use, when you select one item (in a group), a menu will appear and you can choose what to do with the selected item (press or drag).

If you use multiple switches, you can set each switch to perform specific actions and customize your project selection method. Instead of scanning screen objects automatically, you can configure the switch to the next or previous item as needed.

You can change the functionality of the switch controls in a variety of ways to meet your needs and style.

Open "Change Controller" on Apple TV

You can utilize an external Bluetooth® device (such as a keyboard) to connect the switch to an Apple TV, or you can use the platform switch function to transmit switching functionality to other devices (such as an iPhone or Mac). For iOS or iPadOS devices, tap the screen to turn on the switch. (For iPhone 6s and iPhone 6s Plus or later, tap the screen firmly.)

Add a switch and choose its action

- In the Apple TV settings "Settings", go to "Accessibility"> "Change Control"> "Change", select "Install New Change", and follow the instructions on the screen.
If only one button is added, it will be used as a default option.

If you wish to utilize an external switch, you'll need to connect it to the Apple TV first before it appears in the list of possible switches. Follow the directions included with the change. If the switch uses a Bluetooth connection, you need to pair the device with an Apple TV. See Connect Bluetooth devices to Apple TV for further information.

Turn switch control on or off

- In the "Settings" settings on Apple TV, go to "Accessibility" and turn on or off "Change Control".

Use the control switch on the Apple TV

The following are some basic techniques for using the switch.

Select an item

- When an item is highlighted on an Apple TV, start the switch you set as a "Select Item" switch.

When you use a single button, it is used as a "select item" switch automatically.

Act on the selected item

From the control menu that displays when you choose an object, choose a command. The menu layout is determined by how the print function is set up.

- In the Apple TV settings "Settings", go to "Accessibility"> "Change Control"> "Press Behavior" and select an option:
  - Default in: The control menu usually includes a press button and multiple buttons (two dots below). When you are in the moving area of the screen, navigation bars will also appear. To compress a highlighted item, launch the "Select Item" button while the "press" is highlighted. To view more action keys, select More at the bottom of the menu. If you have multiple changes, you can set one provided for compression.
  - Enable auto-compression: To compress an item, do nothing — when the auto-press interval expires (or 0.75 seconds if you have not changed), the item will be automatically clicked. To view the control menu, launch the "Select Item" button before the "Auto Press" expires. The control menu skips the "press" button and goes directly to the full set of function keys.
  - Always press: Press to select the highlighted item without displaying the control menu. Wait until the scanning cycle ends, then press a button to display the control menu.

Scroll the screen

- Select an item in the moving part of the Apple TV screen:
  - Turn off the automatic press: Select the scroll down button (next to the pressed key) in the control menu. Or, to get more scroll options, select "More" and then "Scroll."
  - Enable automatic press: select Scroll from the control menu. If there are too many actions available, you may

need to select more first.
- Press the V V button: select Home from the control menu.

Control media playback

- On Apple TV, select "Media Control" from the Scan menu to play, pause or rewind, or forward.

Dismiss the control menu without choosing an action

Perform one of the following actions on Apple TV:

- Press when the first item is highlighted and all icons in the control menu.
- Select to exit the control menu.

After the number of cycles specified in "Settings" in "Accessibility"> "Change Control"> "Cycle", this menu disappears.

Perform other hardware actions

- On Apple TV, select any item, select the device from the menu that appears, and use the menu to emulate the following actions:
  - Click the menu button for more tasks
  - Press the volume up button
  - Three times the back button (second generation Siri Remote) or menu button (first generation Siri Remote).

On the Apple TV, change the switch control settings

Adjust basic switch control settings

- In the Apple TV settings "Settings", go to "Accessibility"> "Change Control", you can:
  - Insert the switch and specify its function
  - Adjust the recording speed
  - Turn off auto-scan (only if you add the "Move to Next" button)
  - Set scan to pause the first item in the group
  - Select the number of rotation times on the screen before

hiding the switch control
- Set whether to repeat the movement when a button is pressed, and how long you must wait before repeating
- Add another action to the switch by holding down the long button
- Select the media behavior and set the interval to perform the second switch function to display the control menu
- Set whether the switch needs to be shut down and how long before you can accept the switch action
- Allow the controller to override the risks of accidental switching by accident
- Turn on the sounds or read aloud while recording
- Choose which items from the menu and their order
- Select the items that will appear in the "Change Control" menu
- Set items to be collected during object scanning
- Change the color or size of the selecting cursor

Fine-tune switch control settings

- From the control menu, select the following options:
    - Adjust scanning speed
    - Change the position of the control menu
    - Turn the audio on or off
    - Close the group to scan one item at a time
    - Change the color of the cursor

Scan using the switch on the Apple TV

Scanning an item highlights each item or group of items across the screen until you launch the "select item" button. If there are too many items, "Toggle Control" will highlight them in groups. If you select a group, items in that group will continue to be highlighted. If you select a unique item, the scan will stop and a control menu will appear. When you turn on "Change Control" for the first time, object scanning is the default setting.

You can choose from three scanning methods:

- Automatic scanning will automatically highlight items, one by one.
- Manual scanning uses one button to highlight an object and another to activate it.
- One-step scanning is used to change the motion to highlight the highlight from one object to another. If you do nothing later, the highlighted item will be activated.

Select an item or enter a group

1. Watch (or listen) while the item is highlighted.
2. When the item you want to control (or the group that contains the item) is highlighted, start the switch for the item you selected.
3. Use the project management position and select one project to manage.

Back out of a group

- When a dotted line around a group or object is highlighted, launch your select item.

Dismiss the control menu without performing an action

Do one of the following:

- When the item itself is highlighted, launch your select item button.
- Select to exit the control menu.

As things are highlighted, hear their names

Do one of the following:

- In the Apple TV settings "Settings", go to "Accessibility"> "Change Controller", then open "Voice".
- From the control menu, go to "Settings," then "Voice On."
- In the Apple TV settings "Settings", go to "Accessibility"> "Change Control"> "Default Time".

41

Restart, Reset, Update

Restart Apple TV.

If Apple TV does not respond, try restarting.

Restart Apple TV.

Make any of the following:

- Press and hold back button to Siri Remote (Dead of Siri Remote Control) or menu button (first generation of Siri away
- Disconnect the Apple TV from the power output area, wait five seconds, and re-connect.
- To restart your Apple TV, navigate to Settings > Apple TV > System > Restart.

Reset Apple TV.

If you experience problems after restarting and Apple TV has no response, try to reset the Apple TV to factory settings. When you reset the Apple TV, all information is removed, including your account with the configuration.

You can reset the Apple TV and update its software.

Reset Apple TV

1. Open Setting Settings to Apple TV.
2. Go to the program> Reset, and select reset.

Reset Apple TV and return it to factory settings. You can take time, so please wait patiently. If Apple TV is responding, do one of the following:

- If you have Apple TV 4K: Contact Apple TV Support.
- If you have Apple TV HD and Macos Catalina or later: Get the Excel TV cables and HDMI. Connect only one of the USB-C cables (Sold separately) on one end to the Apple TV and your computer. Open Invention on your com-

puter, select Apple TV on Sidebar and select Restore. If this does not work, please contact Apple TV support
- If you have Apple TV HD and PC, you have an iTunes or with Maces Mojave or previously: Disconnect the Apple TV and HDMI Cable Connections. Connect only one of the USB-C cables (Sold separately) on one end to the Apple TV and your computer. Open iTunes on your computer, select Apple TV on Source List and select Restore. If this does not work, please contact Apple TV support.

Reset Apple TV and update software

1. Open Setting Settings to Apple TV.
2. Go to the application> Reset, and select Reset and update.

Select this option to restore Apple TV to factory settings, delete all information settings and update to the latest version of TVOS.

Update Apple TV Software

You will see the message on the Apple TV when you update the available software. You can also view it again to update or set up Apple TV on the default update.

1. Open Setting Settings to Apple TV.
2. Go to the program> Software updates and select Renewal software.

The message is displayed if you can use it.

Check for updates

1. Select Download and enter to start downloading.
2. You can disconnect your Apple TV during the renewal process. The situation is a little bright during the renewal process.

Set automatic updates

1. Open Setting Settings to Apple TV.
2. Go to the program> Software updates, and select Auto Update.

Siri and dictation

Talk to your Apple TV

Siri makes communicating with Apple TV easy, fun, and rich in content. You can look for movies, TV shows, music, or apps, as well as actors and directors you like. Control the playback; access your apps; and inquire about sports, weather, and stock prices — whatever happens on screen.

On Apple TV, Siri will not respond to you as it does on your iPhone or other devices, but it will process your request and display the results on the screen.

See a list of possible questions for Siri

- Press the Siri button on the Siri remote control.

Using your voice, search for and control APPLE TV

- Press and hold the Siri button on the Siri remote control and start talking.

Siri understands the broader questions and applies them to the current context as accurately as possible. For examples of what Siri can do, see Searching for content and apps on Apple TV and other things you can ask Siri on Apple TV.

Dictate instead of type

Whenever you see a text input field, you can use your voice instead of the on-screen keyboard.

- Press and hold the Siri button on the Siri Remote, and speak the text you want to enter. You can also specify a single letter - for example, when entering a username and password.

Use Siri Dictation on Apple TV for further details.

Type Siri queries

You can submit your question to Siri instead of using your voice.

For more information, see Type instead of talking to Siri on Apple TV.

On Apple TV, look for entertainment and apps

Siri on Apple TV can help you find movies and TV shows to watch, and apps you can download from the App Store. Siri can search for movies and TV series based on several factors, including titles, genres, actors, directors, ratings, and age restrictions. You can also request "good" or "popular" results. In apps, you can search by app name, developer, or category.

Search for movies, TV shows, and app

Ask Siri. Say this:

- "What should I watch?"
- Looking for "Hidden People"
- Playing the third episode of the world's first season"
- "Show me pop jokes"
- "What's the best baseball movie?"
- "Want Amy Adams Movie"
- "You want a documentary about cars"
- I was hoping to see a movie in 4K
- "Find my movie in HDR"
- "Watch a beautiful drama"
- "I've discovered several kid-friendly TV shows."
- "Showing PG-13 movies"
- "You want a crossroads"
- "Get the weather app"
- "What are the new sports apps?"
- "Get apps by activating"

Combine search terms

You can also include types of movies and TV shows.

Ask Siri. Say this:

- "Find me some great horror movies"
- "Show me independent foreign movies"
- "Search criminal documents"

After doing your initial search, you can ask more questions to find out what you want to watch. See Configure search results for movies and TV shows on Apple TV.

Search for free content

You can ask Siri to reduce search results in content that can be played faster depending on the content you have and the services you subscribe to.

Ask Siri. Say this:

- "Looking for free children's movies"
- "What are the free movies?"

When searching for a movie or TV show, Siri will search various programs to find exactly what you want. If you request a specific title, Siri will take you directly to the video details page. If there are too many results, Siri will show you the search results to select or continue to improve your search.

Search within a specific app

You can search for YouTube items by installing "YouTube" in the Siri application (you need the YouTube app, which is available in the App Store on Apple TV). Siri will automatically open the YouTube app and add your request to its search screen.

Ask Siri. Say this:

- "Search Coldplay on YouTube"
- "Show me children's art on YouTube"
- "Search Minecraft on YouTube"

You can also search directly on most apps.

For a list of Siri-compatible apps, see the Apple Support article Search on Apple TV.

Search for podcasts

You can directly search for podcasts by adding "podcasts" to your query.

Ask Siri. Say this:

- "Find great podcasts"
- "Get podcasts about surfing"
- Play Radiolab podcast

Once you subscribe to the selected podcast, Apple TV will immediately start playing the current episode.

To learn more about using search results, see Improving search results for movies and TV shows on Apple TV.

Play live content on Apple TV

Some apps on Apple TV include real-time video streaming and content such as news or sports. You can use Siri to directly access these real-time updates.

Note: In some nations or locations, not all features and information are available.

Play live channels

Say the name of the channel you want to watch, and the live video stream will start playing instantly.

Ask Siri. Say this:

- "View ESPN"
- "Open CBS news"

For a list of supported applications by countries/regions, see the Apple Support article Let Siri watch live channels on Apple TV.

Play live sports

Ask Siri about the sports team, points, or schedule. If the game is available in a supported system, Siri will take you directly.

Ask Siri. Say this:

- I'm looking forward to seeing the Warriors play."
- "Which hockey game?"

Select the Live View button on the screen to open the app and watch the game.

You can also request general exercise schedules.

Ask Siri. Don't utter stuff like, "Who's playing baseball tonight?" or "Who's playing baseball tonight?"

Siri will show you a list of all related games, start time and school (once the game has started), and a "Watch Now" button if the supported app is currently playing the game.

Navigate to Apple TV with Siri

Because Siri can detect screen labels, you can use your voice to navigate to any screen on an Apple TV. You may also utilize Siri to make adjustments to your settings.

Use Siri to navigate APPLE TV

- Press and hold the Siri button on the Siri Remote, and say the name of the screen controller, similar to the menu item at the top of the iTunes movie store.

Apple TV prefers control.

Use Siri to change settings

1. Open settings on Apple TV.
2. Press and hold the Siri button on the Siri remote control button and say the name of the setting you see on the screen to turn on or activate the setup.

On Apple TV, you can turn Siri on or off.

You may be prompted if you want to utilize Siri when you first set up an Apple TV (depending on your nation/area and language). At any time, you can alter your mind and turn Siri on or off.

Turn Siri on or off

1. Open settings on Apple TV.
2. Go to "General"> "Siri" and turn off or turn off Siri.

Basics

Use tvOS Control Center on Apple TV

Control Center lets you quickly access settings and functions like switching users and playing music, as well as accessing HomeKit cameras and scenes, putting Apple TV to sleep, and more. You can quickly switch between users in Control Center so everyone has their own Up Next video list, video, and music collection, Game Center data, and recommendations.

Open control center

- On Siri Remote, press and hold the TV button.

Apple TV screen showing the control center

Switch to another user

1. Press and hold the TV button on the Siri remote to open the control center.
2. Choose a user.

When you switch to another user, the previous user closes and the TV and music apps are updated with the new user's unique "playlist", collection of videos or music, and recommended content.

Note: Switching users do not change the Photos app or other settings associated with your iCloud account.

Put apple tv to sleep

1. Press and hold the TV button on the Siri remote to open the control center.
2. Choose to sleep.

For information about changing sleep settings, see Sleep or wake up Apple TV.

Access the current song in music

When a song in the Apple TV Music app is playing or paused, it will be displayed in the control center.

1. Press and hold the TV button on the Siri remote to open the control center.
2. Select the song that is currently playing.

The music app opens on the screen that is playing.

Control music playback on Apple TV by looking at what's now playing.

Access audio controls

1. Press and hold the TV button on the Siri remote to open the control center.
2. Select the audio control button, then select the headset or one or more speakers.

For more information about sharing audio with Apple wireless headphones or playing audio through multiple speakers, see Use Apple TV play audio at home.

Homekit sceneries and cameras can be controlled

In the control center, you can activate HomeKit accessories and view the home cameras you have configured in the "Home" app on iOS, iPadOS, or macOS (Catalina or later) devices connected with the same Apple ID.

For example, you can view video from a compatible security camera and receive a notification when a compatible doorbell camera detects someone at your door. If HomeKit recognizes this person from your photo library, you will receive a personal notification. You can also play the scenes you created in the Home app on iOS, iPadOS, or macOS devices (Catalina or higher). The scene allows you to control various accessories in your home. For example, you can create a TV-watching scene by dimming the lights and adjusting the temperature.

1. Press and hold the TV button on the Siri remote to open the control center.
2. Select the HomeKit button HomeKit button.
3. Do the following:
    - View security camera footage in real-time: Swipe to

swap cameras or select a camera to view the image on full screen.
- Play favorite scene: select scene.

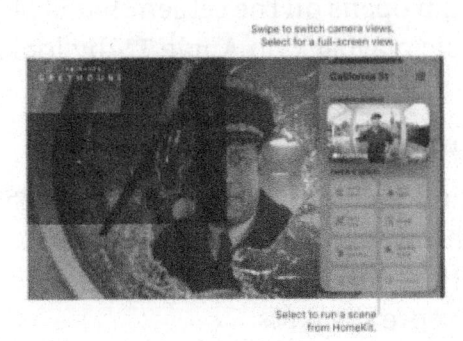

Ask Siri. On Siri Remote, hold the Siri button and say "Show me" in the Home app, followed by the camera name:
- "Show me the front yard camera"

Note: Apple TV cannot control scenes involving compatible HomeKit secure products (such as locks, electric doors or windows, security systems, and garage doors). You must use an iOS, iPadOS, macOS, or watchOS device to control these devices.

To learn more about creating scenes and configuring cameras in the Home app, see the iPhone User Guide.

Quickly access the search app

1. Press and hold the TV button on the Siri remote to open the control center.
2. Select the Search button.

Close control center
- Press the Back button to return (2nd generation Siri Remote) or the menu button (first generation Siri Remote).

Customize the Apple TV home screen

You can change the order of apps on the home screen and put your favorite apps on the top row.

You can also delete apps from the App Store that you no longer use from the home screen, as well as modify the appearance of the backdrop and menus between bright and dark.

Rearrange apps

1. Navigate to an app and press and hold the touchpad (Siri Remote 2nd Generation) or the center of the touch surface (Siri Remote 1st Generation) until the app starts shaking.
2. Drag the application to a new location on the home screen.

Tip: Move your favorite apps to the top row of the home screen; shortcuts to content will appear at the top of the screen when you highlight apps from the top row.

3. To save the new arrangement, press the touchpad (Siri Remote 2nd Generation) or the center of the touch surface (Siri Remote 1st Generation).

Create a folder for apps

You can organize related applications into folders. For example, you can store all music applications in a Music folder.

1. Navigate to an app and press and hold the touchpad (Siri Remote 2nd Generation) or the center of the touch surface (Siri Remote 1st Generation) until the app starts shaking.
2. Drag the application onto another application until a folder appears, then release the trackpad or touch-sensitive surface.
3. To save the new arrangement, press the center of the trackpad or touch the surface.
4. To change the name of the new folder, make sure the folder is highlighted, swipe up, and enter a custom name using the on-screen keyboard or dictation.
5. To close the keyboard:
   - Siri Remote (2nd generation): Press the back button once to close the on-screen keyboard, and then press the back button again to return to the main screen.

- Siri Remote (1st generation): Press the menu button once to close the on-screen keyboard, and then press the menu button again to return to the main screen.

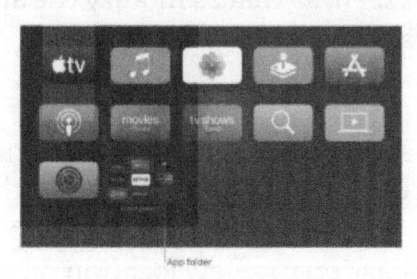

Delete a folder

1. Navigate to the application in the folder, and press and hold the trackpad (Siri Remote 2nd Generation) or the center of the touch surface (Siri Remote 1st Generation) until the app starts shaking.
2. Drag the app back to a row on the home screen, then release the trackpad or touch-sensitive surface.
3. Repeat the above steps for each application in the folder.

When you delete the last application from it, the folder will be deleted.

4. To save the new arrangement, press the center of the trackpad or touch the surface.
5. To return to the home screen, press the back button, back button (Siri Remote 2nd generation), or Menu button (Siri Remote 1st generation).

Move an app into a folder

1. Navigate to an app and press and hold the touchpad (Siri Remote 2nd Generation) or the center of the touch surface (Siri Remote 1st Generation) until the app starts shaking.
2. Drag the application onto the folder, then release the trackpad or touch-sensitive surface.
3. To save the new arrangement, press the center of the trackpad or touch the surface.
4. To return to the home screen, press the back button, back

button (Siri Remote 2nd generation), or Menu button (Siri Remote 1st generation).

Delete an app

1. Select the application you want to delete and press and hold the touchpad (second-generation Siri Remote) or the center of the touch surface (first-generation Siri Remote) until the application starts to shake.
2. Press the play/pause button to view more options, and then select delete.

Deleting an application will also delete its data. You can re-download any application purchased from the App Store for free, but you cannot restore the data.

Change the appearance of apple tv

1. Open the settings on Apple TV.
2. Go to "General"> "Appearance" and select "Light", "Dark" or "Auto".

When set to "Auto", the appearance changes from a light color during the day to dark color at night.

Sync apps and the home screen between several Apple TVs

If you have multiple Apple TVs associated with the same Apple ID, you can maintain the same look and feel for each device. This makes switching between Apple TVs in different rooms, such as the bedroom and the living area, a snap.

1. Open the settings on Apple TV.
2. Go to Users and Accounts> iCloud and enable a home

screen.

Use iOS or iPadOS Control Center to control Apple TV

You can use the Apple TV remote control in the control center of your iOS device or iPadOS. These controls are immediately enabled when you connect an iOS smartphone running iOS 12 or later, or an iPadOS device running iPadOS 13 or later to Apple TV, such as during the setup process or when entering text using the keyboard.

You can use the touch area and buttons to control Apple TV.

Note: Other controls will appear during playback. Click the "Back" button for 10 seconds to go back 10 seconds or click the "Forward" button for 10 seconds to go back 10 seconds.

Add apple tv control to an ios control center

If you don't see the Apple TV Remote icon in the control center of your iOS or iPadOS device, you can add it manually.

1. On an iOS or iPadOS device, go to "Settings" > "Control Center" and tap "Custom Control".
2. Tap the "Add" button and "Add" button next to the Apple TV Remote in the "More Controls" list to add it to the Control Center.

Open apple tv remote control in ios or ipados control center

- On your iOS or iPadOS device, swipe to open the control center and tap the Apple TV remote icon iOS remote control icon.

Navigate and select apps, content, and list

After connecting the Apple TV remote control to the iOS or iPadOS device, do the following:

- Browse apps and content: swipe left, right, up, or down in the touch area.
- On the Apple TV screen, the highlighted app or content item will expand slightly.

As you swipe the touch area, the highlighted app appears larger.

- Navigate the list: Swipe up or down a few times in the tap area to scroll quickly.
- Select an item: swipe to highlight the item and tap the touch area.

View additional menu options

- Press and hold the touch area on the Apple TV remote control on the iOS or iPadOS device after highlighting an item on the Apple TV screen.

If you highlight items with additional options, they will appear in a pop-up menu on the Apple TV screen.

Use the app keyboard

When the onscreen keyboard appears on Apple TV, a keyboard appears on your iOS or iPadOS device's Apple TV remote control.

- Enter text on iOS or iPadOS keyboard.

The text on the Apple TV screen will update as you type.

Adjust the volume

- Press the volume button on the iOS or iPadOS device.

Note: This only applies to compatible audio/video receivers. For information on setting up Apple TV to control your TV or receiver, see Use Siri Remote control TV and volume.

Go back to the previous or home screen

From the Apple TV Remote on iOS or iPadOS devices, do one of the following:

- To return to the previous screen, click here: Select "Menu" from the drop-down menu.
- Return to the home screen: tap and hold the menu button.
- Return to Apple TV App: Tap the TV button once to go to the "Pending List" in the Apple TV App.

Set the TV button destination on Apple TV to change the behavior of the TV button.

View open apps

- On the Apple TV remote control on an iOS or iPadOS device, tap the TV button twice.

See Quickly switch between Apple TV apps.

View open apps

- On the Apple TV remote control on an iOS or iPadOS device, tap and hold the TV button.

On Apple TV, see Using tvOS Control Center.

Open tvos control center

- After opening the Apple TV remote control on your iOS or iPadOS device, press and hold the Siri button.

See Talk to Apple TV.

Activate Siri

- After opening the Apple TV remote control on your iOS or iPadOS device, press and hold the Siri button.

See Talk to Apple TV.

Use the Remote

Siri Remote overview

The Siri Remote included with Apple TV varies by model.

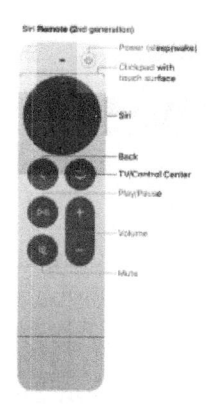

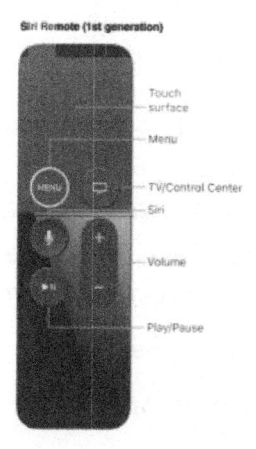

Control what's playing on Apple TV

You have complete control over whether you're watching TV series or movies, or listening to music or podcasts.

Play or pause

- Siri Remote (2nd generation): Press the center of the

touchpad or press the play/pause button to play/pause.
- Siri Remote (1st generation): Press the touch surface or press the play/pause button to play/pause.

When playback is paused, the playback timeline will appear, showing elapsed time and remaining time. Solid bars indicate how many items are cached (temporarily downloaded to Apple TV).

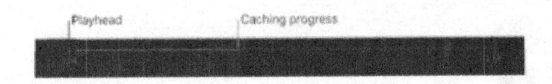

Rewind or fast forward

You can fast-forward or rewind the video in a variety of ways. Use Siri Remote to conduct the following during playback:

- Rewind or fast forward for 10 seconds: Press left or right on the trackpad ring (second generation Siri Remote) or touch surface (first generation Siri Remote). To skip another 10 seconds, press the button once again.
- Rewind or fast forward: Press and hold the trackpad ring (second generation Siri Remote) or the touch surface (first generation Siri Remote) left or right. Press left or right several times to cycle through the fast rewind or fast forward speed options (2x, 3x, 4x). Press the center of the trackpad or the play/pause button to resume playback.

Scrub video backward on the playback timeline

You can browse video thumbnails backward or forwards in the playback timeline to find scenes.

1. Press the play/pause button to play/pause or press the center of the touchpad (second generation Siri Remote) or touch surface (first generation Siri Remote) to pause the video.

At the bottom of the screen, a little preview thumbnail will display above the playback timeline.

2. Do the following:
   - Siri Remote (2nd generation): Swipe left or right to go back or forward in time. Place your finger on the trackpad ring's the outside border until the ring icon displays onscreen, then circle your finger counterclockwise or clockwise around the trackpad ring for more accurate control.

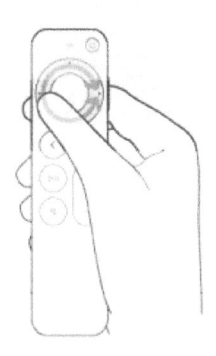

   - Siri Remote (1st generation): Swipe left or right to go back or forward in time.
3. Do the following:
   - Start playback from a new location: press the center of the touchpad or touch-sensitive surface.
   - Cancel and return to original position: Press the return key to return (the second generation Siri Remote) or the menu key (the first generation Siri Remote).

Turn on subtitles and closed captioning(if available)

During playback, you can get more information about the movie or TV show.

- Siri Remote (2nd generation): Press the trackpad ring or swipe down on the trackpad to display the "Information" panel.
- Siri Remote (1st generation): Swipe down on the touch surface to display the "Information" panel.

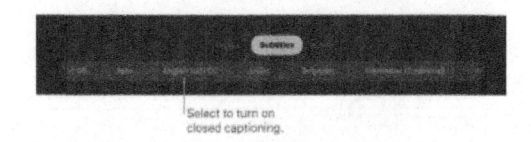

Turn on the picture in picture viewing

When playing or browsing other content on Apple TV, you can play movies or TV shows in a small viewer running in the foreground.

Note that not all programs support picture-in-picture. Contact your app developer to see if it's compatible.

1. In the Apple TV app's TV app or the app of your choice, select a movie or TV show to start playing.
2. During playback, place your finger on the touchpad (second generation Siri Remote) or touch surface (first generation Siri Remote) to display the playback timeline.
3. Swipe or press up on the touchpad or touch surface to highlight the picture-in-picture button that appears on the screen, then press the center of the touchpad or touch surface.

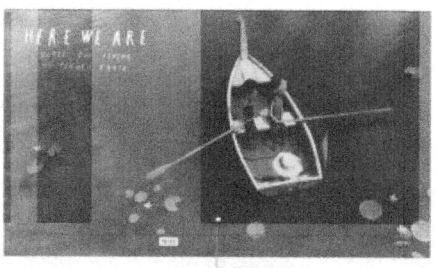

The movie or TV show currently playing will move to a small viewer in the corner of the screen.

    4. In the main area screen, browse and select another TV show or movie to play at the same time.

The new video plays full screen behind the small corner viewer.

    5. During PIP playback, perform any of the following operations:

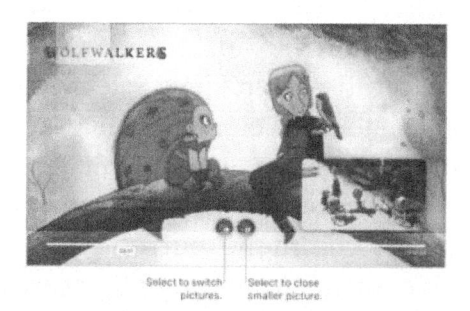

- Swap larger images for smaller images: Place your finger on the Siri Remote trackpad or touch surface to display the playback timeline, then select the "Swap Image in Image" button displayed on the screen.
- Switch audio from large picture to small picture: Press the TV button on Siri remote control. To change the audio back to the large image, press the Back button (Siri Remote 2nd Generation) or the Menu Button (Siri Remote 1st Generation).
- Turn off smaller images: Place your finger on the Siri Remote's trackpad or touch surface to display the playback

timeline, then select the Image off button to turn off the picture-in-picture.
6. To see more options, press the TV button on Siri Remote and do one of the following:

Press the TV button on the remote to see these additional controls.

- Move the viewer to any corner of the screen: select the move screen button "Move" button to reposition the viewer to a new corner. To move it to the next corner, select it again, and so on.
- Return to full-screen view: Select the full-screen button.
- Stop playback and close the viewer: select the close button to close the image in the image.
7. Press the Back button, Back button (2nd generation Siri Remote), or the menu button (first generation Siri Remote) to continue watching in picture-in-picture mode.

Reconnect Siri Remote to Apple TV

Siri Remote will automatically pair with your Apple TV. If it disconnects or you change the remote, you'll need to manually pair it with Apple TV.

Only one Apple TV can be paired with Siri Remote at a time. When pairing the remote, any other paired remotes will be automatically unpaired.

Pair Siri Remote with Apple TV

1. Turn on Apple TV and position Siri Remote 8 to 10 cm

(3 to 4 inches) away and point it toward the front of the Apple TV.
2. Do one of the following:
   - 2nd generation Siri Remote: Press and hold the Back and Volume Up buttons for 2 seconds.
   - Press and hold the menu and volume up buttons for 2 seconds on the Siri Remote (1st generation).

When Siri Remote is successfully paired, a message will appear on the screen.

Quickly switch between Apple TV apps

On Apple TV, you may easily switch between different apps without having to return to the Home screen. Viewing app switching on Apple TV is very similar to multitasking on iPhone or iPad. If the app is not working properly, you can force it to quit and open it again from the main screen.

Switch between apps or force apps to close

Using the Siri Remote, push the TV button twice quickly. Each open application's window is presented in a single line on the screen.

1. In the app switch view, navigate to another app in the center of the screen and do one of the following:
   - Switch to the highlighted app: Press the center of the trackpad (Siri Remote 2nd Generation) or tap the surface (Siri Remote 1st Generation).
   - Force the highlighted app to quit: Swipe up on the trackpad or touch-sensitive surface.
2. Press the Back button, Back button (Siri Remote 2nd gen-

eration), or menu button (Siri Remote 1st generation) to exit the app and move between perspectives without changing the app.).

PODCASTS

Get podcasts on Apple TV

To listen to your favorite podcasts on Apple TV, open the Apple Podcasts app.

Find and play podcasts

1. Open the Podcasts podcasting app on Apple TV.
2. From the menu bar, navigate to any of the following categories:
   - Listen Now: Look for episodes you haven't heard or watched, or continue to play episodes you haven't finished.
   - Browse: Browse selected podcasts.
   - Top Rated: Browse the most popular podcasts by genre.
   - Library: See your podcasts and the list of stations you've created.
   - Search: Enter a search query using the onscreen keyboard.
3. To play or stream an episode, select the podcast and then select an episode.

Use Siri to find podcasts

Do one of the following:

- From the Podcast search screen in the Apple TV Podcast app, navigate the text input field, press and hold the Siri button on Siri's remote, and speak.

For example, say the name of the podcast you are looking for. For more information, see Using Siri Dictation on Apple TV.

- Press and hold the Siri Siri button on the Siri remote anywhere else on Apple TV and type "Podcasts" in the search box.

Ask Siri. Say it like this:
- "Get some interesting podcasts"
- "Find surf podcasts"
- "Play the podcast" Radiolab "

If you followed the selected podcast, Apple TV will immediately start playing the current episode.

Get new episodes as they're released

Follow the podcasts to get new episodes available.

- When navigating to "Featured" or "Top Graphics" in the "Podcasts" app "Podcasts" on Apple TV, select a podcast and then select "Follow".

Delete unplayed episodes

- In the Apple TV podcasting app, navigate to "listen now" or "library", navigate to a single episode, press and hold the center of the trackpad (second generation Siri Remote) or touch surface (first generation Siri Remote) and then select "Delete".

On Apple TV, organize podcasts into radio stations

You may create stations for your favorite podcasts, which will update automatically across all of your devices.

Create a podcast station

1. Open the Podcasts podcasting app on Apple TV.
2. Navigate to Library in the menu bar and select the Add New Station button.
3. Enter your station name and select Done.
4. To access the radio station options, press the Back button, Back button (2nd generation Siri Remote), or menu button (first generation Siri Remote).

5. Update any of the following station options:
   - Station name: If desired, you can enter a different station name.
   - Play: Select the playing order of podcasts on the radio station.
   - Episodes: Select the episodes to be included in the radio station.
   - Choose whether to include audio podcasts, video podcasts, or both types of podcasts.
   - Not Played Only: Enable to only include episodes that you haven't played yet.
   - Select podcasts: Select the watched programs you want to include in the radio station, or select include all podcasts to include all watched programs.

To create other stations, select the "New Station" button and the "Add" button in the library.

After creating a station, select it from the library and then select the station settings to delete it or change its settings.

Browse station episodes

1. Open the Podcasts podcasting app on Apple TV.
2. Navigate to the library and select a station to view the episode list.
3. To play an episode, navigate to the episode and press the center of the trackpad (Siri Remote 2nd generation) or touch-sensitive surface (Siri Remote 1st generation) or the play/pause button play/pause button.

Get more options for a station or podcast

1. Open the Podcasts podcasting app on Apple TV.
2. Navigate to the library and then to a radio station or podcast.
3. Press and hold the center of the touchpad (second generation Siri Remote) or touch-sensitive surface (first generation Siri Remote), then select an option to play the pro-

ject, refresh to see the latest episode, change settings from the project or delete the project.

Control podcast playback on Apple TV

Once the podcast starts playing, it will appear on the "Now Playing" screen. Even if you exit Podcast, it will continue to play, but if you start playing video or audio in other applications, the podcast will stop playing.

You may control playback, navigate to different podcasts, and select more options from the "Now Playing" screen.

Ask Siri. Say: "Play" or "Pause"

Control podcast playback

1. After opening the "Now Playing" screen on Apple TV, press the Back button (Siri Remote 2nd Generation) or the Menu button (Siri Remote 1st Generation) to view other controls.

In the "Now Playing" queue, you'll find podcasts or episodes that are related to the one you're listening to. A timeline will also appear, showing elapsed time and time remaining. When the timeline is active, you can also press the center of the touchpad (second generation Siri Remote) or the touch surface (first generation Siri Remote) to play or pause the podcast.

2. Do the following:
   - Pause or play: Press the play/pause button on Siri Remote to play/pause.
   - Press and hold the trackpad ring (second generation Siri Remote) or the touch surface (first generation Siri Remote) left or right during playback to rewind or fast forward. Press the center of the trackpad or tap the surface to pause and press again to resume playback.
   - Go back to the beginning or skip to the next podcast or episode: Press left or right on the trackpad ring or touch-sensitive surface.
   - Rewind or fast forward 10 seconds: Scroll down to high-

light the play head and press left or right on the trackpad ring or touch surface. To skip another 10 seconds, press the button once again.
- To play different podcasts: Highlight the podcast you want to play and press the center of the trackpad or touch-sensitive surface.
- To get to a certain point in a podcast or episode, swipe down to highlight the play head, then swipe left or right on the timeline to rewind or fast forward.

Stream whats playing to Bluetooth or airplay enabled

To learn more about playing audio through multiple speakers or headphones, see Play audio at home with Apple TV.

1. Do the following:
   - After opening the "Now Playing" screen on Apple TV, navigate up and select the "Audio Control" button "Audio Control" button.
   - Press and hold the TV TV button on the Siri remote to open the control center, then select the audio control button.
2. Select one or more playback destinations.

To learn how to connect a Bluetooth device, see Connect a Bluetooth device to Apple TV. For more information about AirPlay streaming, see Using AirPlay to Stream Audio and Video to Apple TV.

See more options

- After opening the "Now Playing" screen on Apple TV, navigate to the episode in the center of the screen and press and hold the center of the trackpad (second generation Siri Remote) or touchpad (first generation Siri Remote) to see more options:
  - Mark podcasts as played or not played
  - See the full podcast description
  - follow podcasts

Adjust podcast settings on Apple TV

Adjust settings for podcast

1. Open the settings on Apple TV.
2. Go to Applications > Podcasts to do the following:
   - Allow podcasts to update their subscription in the background: Turn on the background app to update.
   - Sync podcasts with your other iOS and iPadOS devices: Enable Sync Podcasts
   - Automatically start playback of the next episode after the end of the current episode: Enable continuous playback.
   - Set default podcasts: Choose how often podcasts check for new episodes that you subscribe to and choose whether you want to keep the episodes after they finish.
   - Use custom colors based on artwork in each podcast: Enable custom colors.

While listening to podcasts, turn on the screen saver

1. Open the settings on Apple TV.
2. Enable "Show during music and podcasts" in "General" > "Screen saver."

Beyond basics

Restrict access to Apple TV content

You can set Apple TV to restrict certain content so that only authorized users can watch, download, or play items in search results or make purchases. This is also commonly known as parental control. You can restrict various content and activities, such as:

- Buy movies, TV shows, and apps
- Make in-app purchases
- Play iTunes movies or TV shows based on content ratings
- Open the app based on age rating
- Play content identified as explicit
- Prevent downloading or reproduction of items identified as explicit content in the search result
- Play multiplayer games on Game Center
- Add friends in-game center
- Change AirPlay or location settings

Note: Restrictions may not apply to third-party applications. To restrict third-party content, adjust the settings for each app within the app or in the "Applications" area of "Settings".

To set or override restrictions, you must enter a password.

Turn on restrictions

1. Open the settings on Apple TV.
2. Go to "General" > "Restrictions", activate restrictions, and enter the 4-digit password.

After entering and verifying the password, restrictions and other options in the restrictions menu will be activated.

Change the passcode

1. Open the settings on Apple TV.\
2. Go to General > Restrictions.
3. Select Change password, enter the current password, then enter the new password.

Configure what types of content are Restricted

1. Open the settings on Apple TV.
2. Go to "General" > "Restrictions" and enter the password as needed.
3. Select options to restrict iTunes Store purchases and rentals, types of content allowed, Game Center settings, and other settings.

You must first enable restrictions to configure restriction settings. To access restricted content, you must enter your password at all times.

Remove all restrictions

You can temporarily remove all restrictions and add them again later.

1. Open the settings on Apple TV.
2. Go to "General" > "Restrictions" and select "Restrictions".
3. Enter the password and then disable the restriction.

If you use "Family Sharing", you can also use iOS or iPadOS devices to limit the content your family members can buy, including enabling "Ask Before You Buy", which allows your child to obtain only parental consent to purchase goods. See Apple TV Family Sharing for further details.

Family Sharing on Apple TV

Family sharing lets you share apps and viewing permissions with up to six family members. An adult in your home (the family organizer) asks family members to join the group and agrees to pay for any purchases made on the iTunes Store or App Store by family members. (Multi-user switching on Apple TV is distinct from family sharing on other Apple devices.)

Important: You cannot turn on or set up Home Sharing directly on Apple TV. Must be done on a Mac, iOS device, or iPadOS device. See the Apple Support article Set up family sharing for additional information on how to do so.

Manage Apple TV Storage

Video and audio are streamed from your Apple TV, but apps are stored locally. The more apps you download to Apple TV, the more storage space you use. At some point, you may receive a warning about insufficient disk space. In that case, you can delete the app to free up space.

Remove unused apps on the home screen

You can set Apple TV to delete unused apps to save space. Deleting the app will keep your documents and data in case you download the app again later.

1. Open the settings on Apple TV.
2. Go to the app and open it to uninstall unused apps.

Reinstall unused apps on the home screen

If you've set Apple TV to uninstall unused apps, they'll appear on the home screen with a download icon.

If the app is still available from the App Store, reinstalling the app will restore your data.

- Navigate to an unused app on the home screen and select it to download the app again.

Manually delete apps to recover space

1. Open the settings on Apple TV.
2. Go to "General" > "Manage Storage".
3. Navigate to the trash can icon for any item in the list and press the center of the trackpad (Siri Remote 2nd generation) or the touch surface (Siri Remote 1st generation)

The app and all its data will be deleted from the device.

You can also delete the app directly from the home screen. For more information, customize the Apple TV Home Screen.

You can redownload an app from the App Store without having to repurchase it if you delete it. See Find Apps in the Apple TV App Store for more information about the App Store.

Check the app;e tv storage levels

1. Open the settings on Apple TV.
2. Go to "General" > "Manage Storage".

The list of apps on Apple TV shows the amount of space used by each item.

Connect a Bluetooth device to Apple TV

Bluetooth® devices may be connected to your Apple TV. such as MFi certified game controllers (made for iPhone, iPod touch, and iPad), Sony PlayStation compatible controllers, Microsoft Xbox compatible controllers, Bluetooth headsets, wireless keyboards, or other accessories.

If Apple AirPods are associated with the same Apple ID, they will automatically connect to Apple TV.

Locate and connect to a Bluetooth device nearby

Before you begin, refer to your device's instructions to set it discoverable.

1. Open the "Settings" setting on Apple TV and go to "Remote Controls and Devices" > "Bluetooth".
2. Select the device from the list.

For more information, see the Apple support article Using Bluetooth accessories with Apple TV.

Adjust game controller settings

If you have a compatible game controller connected using the tasks above, you can assign specific controls to the buttons.

You can also customize controls for a specific application.

1. Open the "Settings" setting on Apple TV and go to "Remote Controls and Devices" > "Bluetooth".
2. Select Game Controller Settings and then select the connected game controller device. If the device is not con-

nected, select Find Controller.
3. Do the following:
   - Customize your game controller: Select Customize, then select Enabled.
   - Customize a specific application for the game controller: Select the application from the list under Application customization.

A list of customizable controls is displayed, with an image of the game controller.

4. Navigate up or down to highlight controls in the list.

When highlighting a control, its corresponding button will also be highlighted on the graphic for reference.

5. Select the desired control and then select all available options for that control.

To restore default settings, scroll to the bottom of the list and select Restore default settings.

Use another remote to control Apple TV

An appropriate TV remote or receiver, a web-based remote for home management systems, or an infrared remote control can all be used to control Apple TV (also known as a universal remote).

If your TV or receiver supports HDMI-CEC, Apple TV will use the HDMI connection to allow it to be controlled by the TV or the receiver's remote. For more information, see the Apple support article Use Siri Remote or Apple TV Remote to control your TV or receiver.

Web-based remotes send signals to Apple TV over the network, so the remote doesn't have to be pointed directly at Apple TV.

To use the infrared remote control, you need to let it learn the signal generated by Siri Remote.

With Apple TV, use a network-based control

You must first connect the web-based remote to the "Home" app on your iOS device or iPadOS before using it for Apple TV's home control system. Verify that the remote is turned on and connected to the internet.

1. On your iOS or iPadOS device, open the "Home" app.
2. Tap the "Add" button, tap "Add accessory" and follow the onscreen instructions.

The 8-digit HomeKit setup code may need to be scanned or entered on the remote controller (or on its box or documentation). You can assign the remote to the room and name it. The name defines how it appears in the Home app and Apple TV, and how you control it with Siri.

Teach an infrared remote to apple tv

1. Open the settings on Apple TV.
2. Select Remote Controls and Devices from the drop-down menu. Follow the onscreen instructions and learn how to use the remote control.

APPLE TV 4K USER GUIDE

Tips and tricks

Basic touchpad control

Compared to the simple aesthetics of the original Apple TV remote, the latest remote is a little more complicated, but once you get the hang of it, it becomes very simple. Of course, the main control element is the touchpad at the top.

Think of it like a MacBook trackpad. You can slide over it; up, down, left, or right. You can touch it by pressing hard or lightly. Clicking on it is how you choose to launch applications, programs, movies, etc. A light click is rarely used, but it can sometimes be helpful.

Remote control button

Like the touchpad, the remote's physical buttons are also easy to use. At the top are the Apple TV menu and buttons. Below them, you have the volume control button, the play/pause button, and the Siri/voice control button. The Apple TV 4K remote control is a little different, but the main concept is the same.

Return to Home Screen... Quickly

Have you ever found yourself clicking on what you want to watch and noticing that you have 30 menus? You're not alone, however instead of repeatedly pressing the menu button on the remote to return to the Apple TV home page (like most people do), simply hit a button and hold it down. The menu button.

This is very obvious; this one is great. Now you can avoid terrible thumb cramps.

Quickly switch apps or eliminate background apps

We're used to watching multitasking or recent app views on a smartphone, but perhaps not on a TV. With Apple TV, the software is loosely iOS-based, so you can load the recently used app preview, which displays thumbnails/stacked cards of apps you've used before or recently.

All you have to do is double click the TV button and you will start this preview. To select the app to switch to, just swipe the touch-sensitive control panel on top of the remote.

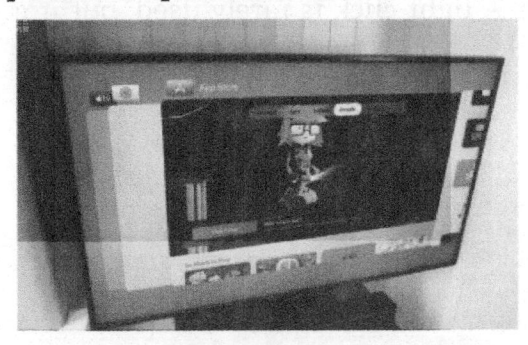

If the app stops responding or experiences problems, this is a very convenient way to close the app before restarting it. Just open the recent app view, swipe until you reach the app, then slide the slider up to remove it from the screen and close it.

Find things with your voice

In the latest generations of Apple TV, Siri has been integrated and can use your voice to request movies or TV shows. All you have to do is press the little button on the remote that looks like a microphone. Press and hold while making a request, then release, you'll see the text appear on the screen in real-time, followed by the relevant results.

You can say something like "Play Spider-Man Heroes Return" or when watching a certain movie or show, you can ask who is in the scene or who directed the movie, and the result screen will be dis-

APPLE TV 4K USER GUIDE

played in a pop-up graphic background.

It also doesn't just apply to content on Apple's iTunes service. You can use Siri to search many other content providers so you can watch it even if you don't own it (as long as it can be streamed elsewhere).

Move, delete, or sort apps

Select the app icon you want to move and press and hold the remote control selection key, suddenly you will see the icon vibrate (similar to how the app icon vibrates in iOS), which means you can move to the left, right, and raise it, or lower the remote. Just use the touchpad and press the select button again to set the content in place. simple.

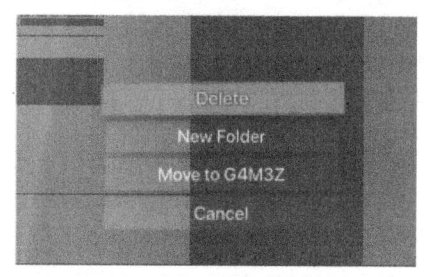

Like iPhone, if you want to create a folder with lots of apps or folders to organize your home screen, you can do the same. Just hover your mouse over an icon on top of another icon to create a folder automatically.

If you just want to delete the app completely, you can delete it. Select what you want to delete, press and hold the select / touchpad button on the remote control until the icon starts to vibrate, and press play/pause to access the new menu. From there, click "Remove".

Take the remote (no app)

One of the most annoying things about any home entertainment device is the remote control, especially when you need to use it to enter text queries. If you're one of the few people in the world - if more than that - people will never lose their remote control, yet

81

you'll still have to endure the experience of manually clicking on each letter while searching.

But like most things in life, there is an app that can do that.

If you've lost your Apple TV remote or just want to use the keyboard to enter queries, download the free Remote app on your iPhone, iPod Touch, iPad, or even (you can believe it) Apple Watch. This is a beautiful and simple thing (and almost impossible to miss).

When you set it up for the first time, it will look for the Apple TV box on your network, and after selecting the box you want to use, enter a code, it will appear on the screen, and then bang With a bang, you finished.

Scrub (find that scene)

If you're looking for a specific part of the movie or miss something while leaving the room to get it (you forget what it is), you can navigate the video timeline on the touchpad by tapping the pause button and swiping left until you find the scene you want. When you swipe, it displays preview thumbnails above the real-time updated timeline, so it's very accurate.

If you prefer an old school, you can rewind or fast forward in 10-second increments by simply clicking on the left or left side of the touchpad.

Subtitles, please.

This is convenient for those who are hard of hearing or just want to watch a foreign movie, but still understand what's going on:

Apple TV lets you enable subtitles. There are several ways to do this, but the easiest way is to swipe down while watching a show.

You will then see the option to allow subtitles to drop at the top of the screen (if available). For example, Daredevil on Netflix lets you choose between English, French, German and Spanish. Amazing.

Dark mode baby

Like all other operating systems in the world today, you can set a dark theme on Apple TV. By default it is set as a light theme, so go to "Settings">"General">"Appearance" and select the "Dark" option, or you can choose "Automatic" to automatically switch at night.

Find out where the amazing live wallpaper was filmed

One of the coolest things about owning an Apple TV is the slow-motion landscape recording on the live wallpaper/screensaver that appears after a few minutes of activity. If you're not sure where you are, simply press the touchpad on the remote and a text will appear on the screen telling you where you are.

Manually activate the live wallpaper screensaver

Normally, when the TV is idle for a few minutes, you will only see those amazing live wallpaper scenes. However, you can activate it manually. Go to the home screen and press the menu button again, it will enter screen saver mode.

Rename your Apple TV

You can rename Apple TV. Yes, who knows? This is useful if you have more than one Apple TV at home and want to know which

Apple TV is on your network at any given time. Navigate to "Settings" > "General" > "About" > "Name" to give your Apple TV a name. You can choose from a list of conventional Apple names at this time.

Apple lets you name your Apple TV based on the room (such as the living room) where the Apple TV is located, but if you want to be more creative, you can specify a custom name. We call our Bob. Hi Bob!

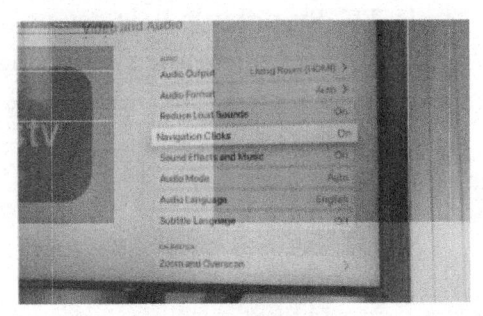

Adjust video quality

If you want to adjust video and audio quality, just go to "Settings" > "Video and Audio", you will find many options here. You can change the resolution or match the TV frame rate with the content.

You can even choose to turn off these navigation "clicks" so that as you scroll through lists and app icons, there's no weird trembling every time you swipe your finger across the touchpad.

The PIN protects what you don't want your kids to see

One of the most useful options on the Apple TV system is the ability to add a PIN code to content with a specific (or higher) rating so that your child doesn't accidentally watch any content you deem inappropriate. The only downside to this native option is that it only works with iTunes content: movies and TV shows you buy or rent through Apple.

To set a PIN code, go to "Settings" > "General" > "Restrictions". After opening it, it will ask you to set a PIN code. You can choose which music or podcasts can be played and establish ratings for movies and TV episodes by scrolling down the list to "Allowed Content."

If your child can watch level 12 content, you can select it as the minimum requirement, and then every time they try to watch a level 15 or 18 movie, you will be asked to enter the PIN code you set previously. For other services (like Amazon Prime), you'll need to go through the provider's own parental control options.

Pairing a Bluetooth headset, game controller, or keyboard

If you want to watch your favorite shows without disturbing other people in the house, you can pair some headphones with Apple TV. Open the settings menu, go to "Remote Control & Devices" and scroll down to "Bluetooth". You will now see a list of available devices.

If you already have AirPods or Beats headphones with an Apple H1 chip, your headphones will have been paired to the TV through iCloud magic. For any other devices, put them into pairing mode and wait for them to appear under "Other Devices".

The same method also applies to Bluetooth keyboards and game controllers. Each controller has its method for putting it into pairing mode. For example, with an Xbox wireless controller, you only need to press and hold the connect button for a few seconds. With PlayStation DualShock 4, you can press and hold the PS and Share buttons at the same time until the indicator light starts flashing.

Sent to sleep

Putting the Apple TV to sleep still has a long way to go, and there's also a short way to go. Of course, you can navigate the settings menu by navigating to Settings > Sleep Now. However, the easiest way is to hold down the Home / Apple TV button until you see a narrow side menu swiping right across the screen.

Here you will see a big button with "Sleep" written on it. After pressing, Apple TV will turn off and your TV (and all other connected devices) will go into sleep mode.

Made in the USA
Las Vegas, NV
01 November 2024